TIME FOR KIDS READERS

The Surprising Mr. Birdseye

by Roberta Ann Cruise

Harcourt

Orlando   Austin   Chicago   New York   Toronto   London   San Diego

Visit *The Learning Site!*
www.harcourtschool.com

MR. BIRDSEYE

BORN: December 9, 1886; Brooklyn, New York

DIED: October 7, 1956; New York, New York

PARENTS: Clarence Frank Birdseye and Ada Underwood

EDUCATION: Amherst College, Amherst, Massachusetts; Class of 1910 (did not graduate)

MARRIED: August 21, 1915, to Eleanor Gannett

CHILDREN: Kellogg, Ruth, Eleanor, Henry

KNOWN FOR: developing quick-freeze method for preserving vegetables, fruits, meat, fish, and other foods. His work led to a new era for frozen foods throughout the United States.

INVENTIONS: He held patents in the United States and Europe for about 300 creations, including an infrared heat lamp designed to keep cooked foods hot.

"I do not consider myself a remarkable person. I am just a guy with a very large bump of curiosity...."
— CLARENCE BIRDSEYE

Anyone who has ever walked through the freezer aisles of a supermarket knows the name Birds Eye. The brand has been linked to frozen foods for more than 70 years. But did you know that there was a real man named Clarence Birdseye? In his time, the idea of quick-frozen foods seemed revolutionary. "Why would we need frozen foods?" people asked. "Where would we store them?" Clarence Birdseye had the answers. He was an inventor whose many interests included plants, animals, and food.

Birdseye is an unusual family name. Clarence Birdseye once explained that the name dates back to England about 500 years ago. A page, or a young servant to a royal person, was asked to shoot a bird for the queen. Birdseye said: "This page-boy ancestor of mine, according to the records, took out his trusty bow and arrow and shot that bird right in the eye. The queen was so tickled she gave him the name [Birdseye] right on the spot." The family name Birdseye soon had a motto that went with it: Stay Right on the Target.

Clarence Birdseye was born on December 9, 1886, in Brooklyn, New York. He died almost 70 years later, on October 7, 1956, in New York City. During his long and interesting life, Birdseye received patents for some 300 inventions. He developed methods for drying foods. He also developed the infrared heat lamps that keep food hot at restaurants.

In the later years of his life, Birdseye found a way to remove water from food to preserve it. These dehydrated foods are very compact and can be packaged in extremely small containers. Hikers and astronauts are among the people who benefited from that invention. Birdseye had many other ideas and interests, but it was his work on what he called *quick-freeze ways to preserve food* that made him both rich and famous. The surprising Mr. Birdseye became known as the father of frozen food.

Clarence Birdseye was born into a family of serious and creative thinkers. His father, Clarence Frank Birdseye, was a lawyer and a judge on the New York State Supreme Court. His mother, Ada Underwood, was the daughter of A. D. Underwood, an inventor. Even as a boy growing up in Montclair, New Jersey, young Birdseye was interested in the natural world. He wanted to learn all about animals as well as plants. His early interests later paired with an interest in food that became stronger when Birdseye enrolled in cooking classes.

The Brooklyn Bridge connects Manhattan and Brooklyn, where Birdseye was born.

Ice fishing gave Birdseye the idea for quick-freezing foods.

Education was valued within Birdseye's family. But there was not always enough money to pay for schooling. After completing high school, Birdseye entered Amherst College in Massachusetts. During his years there, he held various jobs to help pay for his education. Birdseye hoped to study biology, the science of living things, at Amherst. But he left the college because the money to pay for his education had run out.

Even though he didn't graduate from college, Birdseye never lost interest in science or in building his own business. After his college years, he worked as a naturalist with the U.S. Biological Survey, part of the Department of Agriculture. His government jobs included researching a disease called Rocky Mountain spotted fever in Montana.

In 1912 Birdseye was working in Labrador, Canada, in the Arctic Circle. He spent several years traveling back and forth between the United States and Labrador. In Labrador, he studied how Inuits lived in the Arctic region. While watching them ice fish, Birdseye noticed that fish froze rapidly when exposed to extremely cold temperatures. He also discovered that when quickly-frozen fish is later cooked, it has the taste and texture of fresh fish. Later, Birdseye would realize that this method also helps the food keep its nutritional value.

In August 1915, during one of his trips home to the United States, Birdseye married Eleanor Gannett. The next year, in 1916, Mr. and Mrs. Birdseye and their infant son traveled to Labrador. Birdseye was concerned about how they would manage to eat healthful foods during the bitter Arctic winter. At that time, fresh vegetables, fruits, and meat would be scarce. The food Birdseye brought from the United States included several barrels of fresh cabbages. Birdseye placed the heads of cabbage in salt water and then let them freeze quickly in the cold Arctic winds. He also froze other fresh vegetables and enough ducks, rabbits, and caribou meat to feed his family during the long winter.

During the early 1900s, most homes had iceboxes, which were large wooden boxes cooled by a hunk of ice. They could keep food fresh for short amounts of time. In 1911 the first refrigerators for home use were introduced. They were very small, with limited storage space. Their freezer sections were tiny. When the freezers became clogged with ice, as they often did, they had to be chipped out by hand. Few people could keep leftovers, much less frozen foods.

Birdseye's experiments with quick-freeze methods to preserve food started a new industry. His careful research and sharp observation proved the benefits of quickly freezing vegetables and meats. Excited and even more determined, Birdseye learned the importance of patience. It took several years of hard work for Birdseye to turn his surprising discovery into a business.

Birdseye and his family returned to the United States in 1916. He worked at several government jobs until 1922. Then once again he began to experiment with ways to freeze fish. This time he was working in New York City, near the Fulton Fish Market, where much of the fish coming into the area was processed, or skinned and prepared for sale. He spent just $7 to get started. The money paid for the electric fan, ice, and brine, or saltwater, that he needed to quick-freeze fish fillets.

Birdseye made an important change in the methods used to freeze food. He packed fish and other food in cardboard boxes. He then placed the containers between two flat canvas belts and quickly froze the boxed food under high pressure. "My first quick-freezing trials were with fish and rabbits, and I packed them in old candy boxes," Birdseye once recalled. Eventually he

Frozen food couldn't be sold until refrigerators with freezers replaced ice boxes like this one.

Birdseye used fish from the Fulton Fish Market in New York for his quick-freezing experiments.

figured out that flat metal plates worked even better than canvas belts for freezing food. Birdseye's surprising idea was on its way to becoming a reality. Consumers could buy frozen food packaged to take home from grocery stores in cardboard boxes. This was revolutionary!

The small business grew. Even though Birdseye's business was growing it ran into money problems. In 1923 he was forced to go out of business, but he didn't give up. He looked for and found people who would build his business with him. His new partners included several wealthy people who were eager to support Birdseye's ideas. With almost $400,000 to back him, Birdseye started over again in 1924.

Birdseye's new General Seafoods Company was based in Gloucester (GLAWS•ter), Massachusetts. Because it was located on the northeastern coast of the United States, Gloucester had a constant supply of fresh fish. Even though Birdseye now had the money to back his ideas, he had several difficult years before he could make his business work. Even when he wanted to sell his business, no one was interested in buying it. So he had to stick with it!

One day a wealthy woman named Marjorie Merriweather Post stopped at Gloucester aboard her yacht. She had been impressed by a goose that the yacht's chef had prepared for dinner. Post owned the Postum Company, a big and thriving food business. Its directors included her husband, Edward F. Hutton, a banker. When Post discovered that the delicious, tender goose she had enjoyed so much had been frozen at a plant in Gloucester, she went ashore with her husband to visit the plant. They met Clarence Birdseye, who ran the plant. Although Post was interested in Birdseye's idea, she could not persuade Hutton to take frozen foods seriously. He wouldn't invest his money in a product that didn't seem like a sure thing. Hutton's decision turned out to be unfortunate for him. At that time, Postum probably could have bought Birdseye's business for about $2 million—a good deal even then.

In 1927 General Seafoods Company froze more than 900,000 pounds (408,233 kg) of fish. The company also experimented with quick-freezing meat, poultry, vegetables, and fruits. It changed its name to General Foods Corporation. Two years later, in June 1929, the Postum Company and Goldman Sachs Trading Corporation together bought General Foods Corporation for $22 million. Post finally got the frozen-foods business that had caught her attention two years before. At last, Clarence Birdseye's idea made him wealthy! Even though Birdseye may have started the business, it was mainly owned by his partners. They received most of the $22 million. Still, Birdseye never again had to worry about money.

The big sale didn't end Birdseye's involvement with frozen foods. It just marked another beginning. Birdseye was placed in charge of research for a new product line, Birds Eye Frosted Foods. He would work with more than a dozen technicians and researchers. Their goal was to freeze vegetables in a way that would seal in their flavor and color. Together, Birdseye and his Frosted Foods colleagues figured out how to blanch, or quickly boil, vegetables before freezing and packaging them. Birdseye and the others were definitely on to something big. Vegetables that had been blanched, then quickly cooled, frozen, and packaged looked and tasted like fresh vegetables. It wouldn't be long before customers would be filling up their freezers with crisp, fresh-tasting frozen vegetables.

Marjorie Merriweather Post
1887–1973

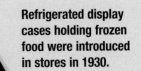

BIRDSEYE
FROSTED FOODS

BEDDING STORES
P AND REST PRODUCTS

FROSTED
FOODS

CLEAN

ALWAYS
FRESH

STANDARD
QUALITY

CONVENIENT
TO BUY AND USE

BEDDING STORES
SLEEP AND REST PRODUCTS

REFRIGERATOR

50°Below Zero

Refrigerated display cases holding frozen food were introduced in stores in 1930.

March 6, 1930, was an important day in the world of frozen foods. On that day, 10 grocery stores in Springfield, Massachusetts, introduced 26 different kinds of frozen foods. An advertisement in the March 6, 1930, edition of the *Springfield Union* talked about a revolution in freezing. Workers from Birds Eye Frosted Foods gave out samples and explained how to prepare them.

Grocery shelves soon featured frozen peas and spinach, raspberries and cherries, fish and beef. Unfortunately, Birdseye's idea found its way into grocery stores just as the U.S. economy sank into the Great Depression. People didn't have extra money to spend on prepared foods. The test in Springfield continued for two years. More testing followed in Rochester, New York, and in Washington, D.C.

In March 1930 a typical grocery store in that city sold about $104 worth of frozen food on a Saturday. Two months later, sales of frozen foods in the same grocery store accounted for more than $243. At first frozen meat was the big seller for Birds Eye. By 1936, frozen vegetables and fruits took the lead in sales at grocery stores throughout the United States.

By 1937 the number of kinds of Birds Eye frozen foods had more than doubled. The 53 choices ranged from asparagus and Brussels sprouts to rump steak and swordfish. Frozen vegetables, fruits, meat, and fish that tasted good and that were easy to buy, store, and prepare made Birds Eye Frosted Foods a success. Advertising and good word-of-mouth helped build sales. So did efforts to educate grocers and consumers about the benefits of frozen foods.

Birds Eye's first advertisement in a magazine appeared in January, 1936. The ad showed drawings of well-dressed people and the headline: "Have you heard the incredible news about GREEN PEAS?" The company guarantee promised that customers unhappy with the product would get their money back. In 1940, the first full-color ad for frozen food appeared in *Life* magazine. That year, American shoppers gobbled up $150 million worth of frozen foods.

Customers crowd one of the 18 stores in Springfield, Massachusetts, that were test-marketing Birds Eye Frosted Foods.

During World War II, from 1941 to 1945, about 19 million women went to work outside the home for the first time. They certainly welcomed time-savers like frozen foods. Convenience foods were no longer a luxury—they were a necessity!

Many people enjoy the convenience of TV dinners.

By the 1950s, another invention was changing the lives of people in the United States. More and more U.S. homes got televisions. It became a common sight to see entire families eating their dinners in front of a flickering TV screen. It was no surprise that, in 1954, the first TV dinner came on the market. By the end of the 1950s, frozen foods were a billion-dollar business in the United States.

As for Clarence Birdseye, he had stayed with Birds Eye Frosted Foods as its president until 1934. He next turned his attention to keeping cooked foods hot. From 1935 to 1938, he was president of Birdseye Electric Company. Anyone who has ever eaten at a fast-food restaurant has benefited from the work Birdseye did to develop infrared heat lamps.

Toward the end of his life, in the late 1940s and early 1950s, Birdseye worked mainly on finding ways to remove the water contained in various foods. Water-free foods could be packaged in handy containers small enough to carry in the pocket of a jacket or a pair of pants. Hikers and astronauts were among the customers for that invention. In 1951 Birdseye and his wife, Eleanor, together wrote *Growing Woodland Plants*, a small publication that tied into his lifelong interest in plants.

From 1953 to 1955, he worked in the South American country of Peru. In Peru, Birdseye put his strong skills as a researcher and creative thinker to use. He worked to develop a method for turning straw, sugarcane, and other raw materials into pulp that could be used to make paper.

In Peru, high up in the mountains, the air was thin. It was often hard to breathe. Clarence Birdseye became ill while working there and suffered a heart attack. He died in New York City on October 7, 1956, as a result of heart problems.

A Quick Look at Preserving Food

Many ripe fruits and vegetables begin to spoil soon after they are harvested. Unpreserved poultry, fish, and meat can also spoil quickly if left unrefrigerated. Food spoils as the cells in plant and animal tissues change. When food spoils, its texture may change and it may lose its nutritional qualities. The rotten food may also begin to smell peculiar, and it is likely to have an unpleasant taste. Humans who eat rotten food can become seriously ill. *Food poisoning* is a term to describe the illness brought on by eating spoiled food. Food poisoning is caused by certain types of bacteria, or germs, that grow in unpreserved food.

Freezing food to preserve it dates back to prehistoric times. Sir Francis Bacon, an English philosopher, is credited with developing the first artificial freezing method. In 1626 he wrote about his efforts to preserve fruits and meats by packing them in shaved ice.

Pasteurizing is the opposite of freezing. It uses high heat to kill bacteria in milk, cream, fruit juices and other liquids. Many foods also use chemicals, especially nitrites and sulphites, as preservatives. These chemicals kill bacteria or slow their growth.

Another way to preserve food is to reduce or eliminate the liquid in it. Bacteria have trouble growing in food that is dried out. Dehydrated foods, such as powdered milk, have no water at all. Salting preserves food in a similar way. In this method, meats and other foods are packed in salt to draw out water, so bacteria can't multiply easily in them.

Today, shoppers can eat fresh-tasting food all year round thanks to Birdseye's freezing process.

Perhaps the most surprising thing about Birdseye's life is the thread that linked his childhood interests to the important work he would do as an adult. The surprising Mr. Birdseye put his lifelong interest in plants, animals, and preparing food to good uses that benefited people around the world. In 1941, more than 30 years after he had left the school, Amherst College honored Birdseye by giving him an honorary degree.

Clarence Birdseye has no statues or buildings to honor him. Instead, his monuments are found in the refrigerated cases that line the walls of supermarkets around the globe!